Introduction To IPTV Billing

Event Recording, Usage Rating, Content License Fees, and Advertising Revenues

Lawrence Harte, Avi Ofrane

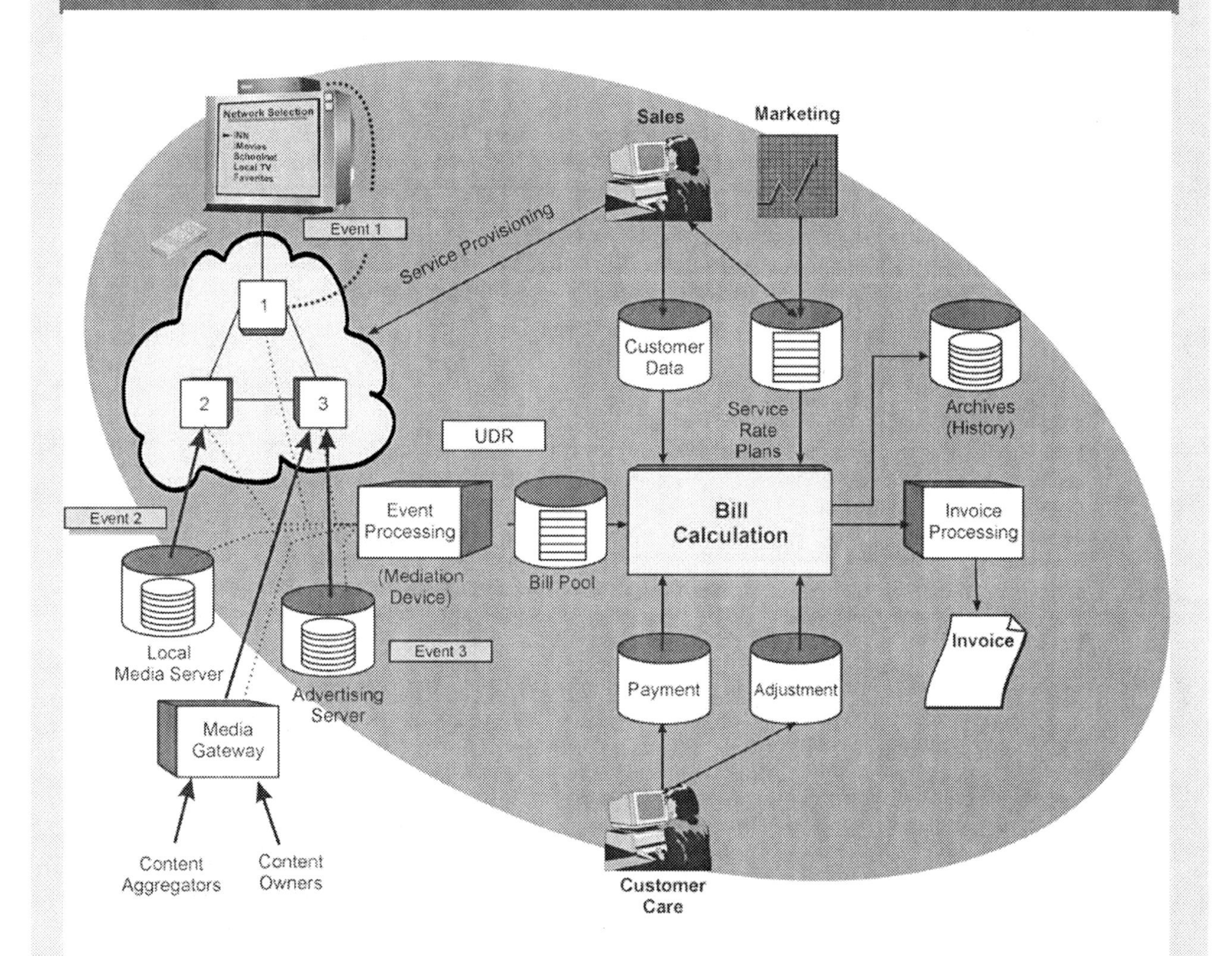

Excerpted From:

Billing Systems

With Updated Information

ALTHOS Publishing

ALTHOS Publishing

ISBN: 1-932813-73-X,

ALTHOS electronic books (ebooks) and images are available for use in educational, promotional materials, training programs, and other uses. For more information about using ALTHOS ebooks and images, please contact Karen Bunn at kbunn@Althos.com or (919) 557-2260

Terms of Use

About the Authors

Mr. Harte is the president of Althos, an expert information provider which researches, trains, and publishes on technology and business industries. He has over 29 years of technology analysis, development, implementation, and business management experience. Mr. Harte has worked for leading companies including Ericsson/General Electric, Audiovox/Toshiba and Westinghouse and has consulted for hundreds of other companies. Mr. Harte continually researches, analyzes, and tests new communication technologies, applications, and services. He has authored over 50 books on telecommunications technologies and business systems covering topics such as mobile telephone systems, data communications, voice over data networks, broadband, prepaid services, billing systems, sales, and Internet marketing. Mr. Harte holds many degrees and certificates including an Executive MBA from Wake Forest University (1995) and a BSET from the University of the State of New York, (1990).

Avi Ofrane founded the Billing College in 1996, a training company addressing the converging market trends associated with telecommunications Billing and Customer Care. The Billing College is a spin-off company of Mr. Ofrane's technology consulting company, Jupiter Data, Inc., established in 1990. Mr. Ofrane began his career in 1977 as an analyst with the IBM Corporation and has since 1982 concentrated exclusively on the telecommunications industry, in which he is now a recognized expert in Billing and Customer Care. Throughout his extensive career, Mr. Ofrane has been involved in all aspects of the industry, from strategic planning and executive management to vendor evaluation and project implementation. Mr. Ofrane lectures extensively on Billing and Customer Care issues, strategies, methodologies, and practices. He is a frequently requested speaker at major North American and European conferences. Mr. Ofrane is currently President and CEO of the Billing College, as well as a master instructor of the company's courses. Mr. Ofrane is the co-author of the book "Telecom Made Simple" and has written numerous articles for international trade publications. Mr. Ofrane holds a Bachelor of Science in Computer Science from Pennsylvania State University.

Table of Contents

Introduction to IPTV Billing

Billing and Customer Care platforms for IPTV systems convert the bits and bytes of transmitted information such as IP Video or other services that are provided within multimedia distribution networks into the money that will be received by the service provider. To accomplish this, the IPTV billing and customer care system provides account activation and tracking, service feature selection, selection of billing rates for specific network events, invoice creation, payment processing, and management of communication with the customer. IPTV billing may combine digital television, telephone, data transmission and information services into a single billing system.

Billing and customer care systems are the link between end users and the communications network equipment. IPTV service providers manage networks, setting them up (provisioning) to allow customers to transfer information, and bill end users for their use of the system. Customers who desire IPTV services select carriers by evaluating service and equipment costs, reviewing the reliability of the network, and comparing how specific services (features) match their communication and media needs. Because most network operations have access to systems with the same technology, the billing and customer care system is a key criterion used to differentiate one service provider from another. IPTV systems can be managed networks or they can be unmanaged.

Managed IPTV

Managed IPTV is the delivery of IP television services over a managed (controlled) broadband access network. Managed IPTV systems can control and guarantee the quality of television services. Managed IPTV systems are traditionally provided by telephone (telco) or cable service providers.

Figure 1.1 shows a managed IPTV system. This diagram shows that for managed IPTV systems, the service provider can control transmission parameters in the user's access network to guarantee a specific quality of service (QoS) level in the access portion of the network. This example shows that a router in a data communications network can be programmed (managed) to give priority to packets with a particular address or type of service. When packets from a high priority source are received (media source #1), they are

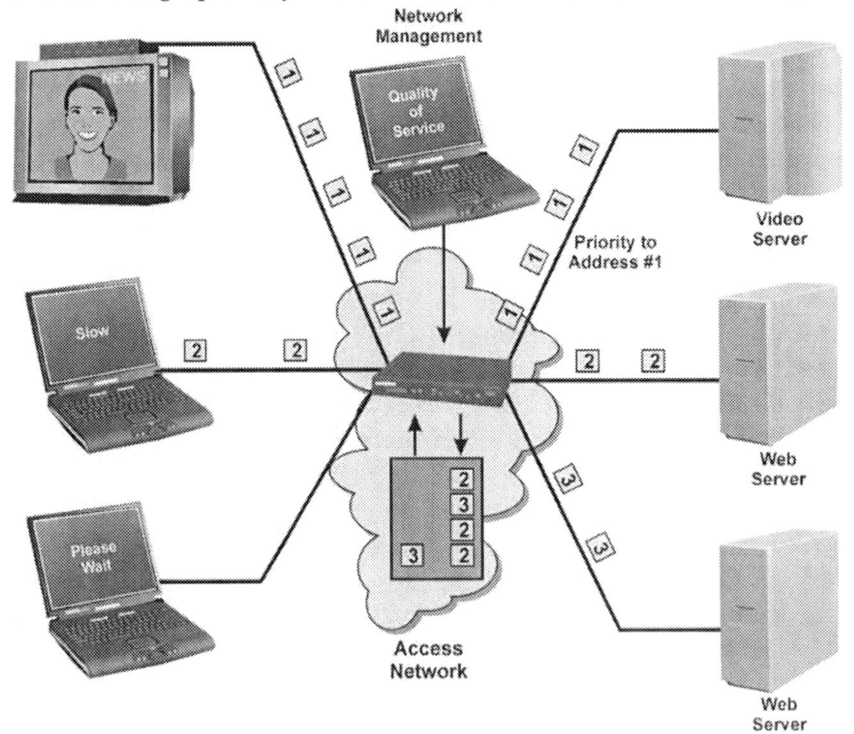

Figure 1.1, Managed IPTV System

given priority over the packets that are received from other media sources. When packets of a lower priority source are received, (media sources #2 or #3), they are discarded or delayed. This example shows that 3 users have different QoS levels. The IPTV customer receives a high priority level ensuring that they obtain a video signal with minimal distortion while the others are provided with data transmission rates that are determined by the remaining capacity (bandwidth) of the access network.

Unmanaged IPTV

Unmanaged IPTV (also called Broadband television) is the delivery of digital television services over broadband data networks. Unmanaged IPTV systems may be able to control and guarantee the quality of television services if the underlying broadband connections have enough bandwidth. Internet service providers or media management companies through broadband Internet connections usually provide unmanaged IPTV systems.

Figure 1.2 shows an unmanaged IPTV system. This diagram shows that the IPTV service company can provide IPTV service to any customer that is attached to a broadband distribution network. This example shows how a company that is connected to a IP broadband data network (such as the Internet) can provide IP television services directly to viewers without being able to directly control the underlying IP network. This example shows that while the unmanaged IPTV service provider may not be able to guarantee the quality of service (QoS), if the bandwidth in each leg of the communication link is high enough, it may be possible to provide traditional television quality.

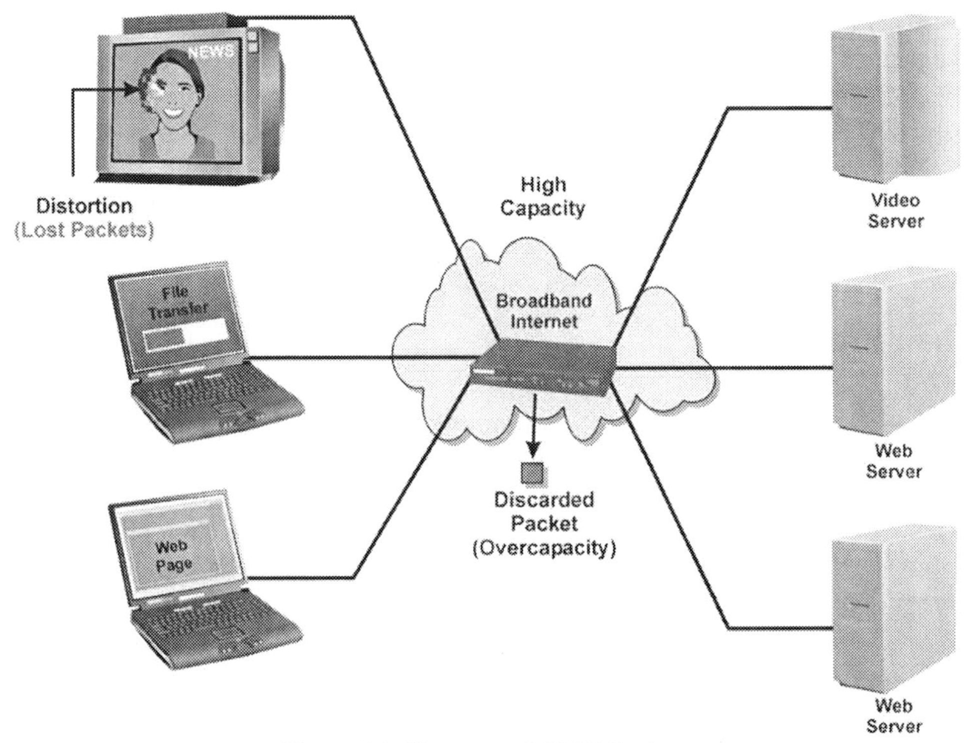

Figure 1.2, Unmanaged IPTV System

Services

IPTV systems typically combine several different types of services. These include television (TV), data communication (e.g. Internet), telephone (voice) and information services (data manipulation).

Billing systems process the multiple usage events of network equipment during the media usage (events) and access/transfer of media into a single usage detail record (UDR). The billing process consists of: receiving event records from various networks and systems, combining multiple events related to a specific service usage into a usage detail record (UDR), determining the rates associated with each record, calculating the cost for each billing record, aggregating these records periodically to produce invoices, sending invoices to the customer, and recording payments received from the customer.

The types of IPTV system parts that a customer may use in a network include system access (basic information transfer), information processing (such as interactive gaming), and content delivery (such as viewing a movie). When the communication service involves system access through different networks, the media is normally routed through a content aggregator and license fees or usage royalty charges may apply. License fees or royalties are the charges for access or use of media content.

Examples of system access services include digital subscriber line, cable modem, wireless broadband, optical fiber and other data connectivity services that transfer information between points. The IPTV service provider may own or control these access networks and bill for data services.

Information processing services include interactive gaming, voice mail, fax store and forward, and other services that involve the processing of information that is passed between two or more points.

Content delivery involves linking customers to sources of information content and transferring the content to the end customer. Examples of content delivery include downloading movies, personal media channels, and the delivery of other sources of media that the customer requests.

Television

Television broadcasting is the transmission of video and audio that is intended for general reception by the public, funded by commercials, subscription services, or government agencies. Traditional television radio broadcasters transmit at high power levels from several hundred foot high towers. While a high power television broadcast station can reach over 50 miles, IPTV can reach anywhere on the globe where there are broadband connections.

IP Television services include the delivery of local content and channels, network programming, on demand content and high definition television channels. Customers typically pay for television services through a combination of basic service fees, subscription channel access fees, and pay per view fees. The basic service fees typically include access to a limited number of television channels (local and network). Subscription channel fees provide the viewer with access to premium channels that are not available in the basic service. Additional usage fees may be charged for specific media selections and viewing events the user consumes over the basic service period.

In addition to the charging for basic content services, advertisers may pay the content provider a fee for the insertion of advertising messages. Because IPTV has the ability to deliver advertising messages to specific customers (addressable advertising), the selection and delivery of advertising messages needs to be tracked and billed to advertisers or advertising agencies.

Figure 1.3 shows typical television services that create billing and service usage records for subscription and advertising services. This diagram shows that television subscription services include access to local content, live network programming and pay per view (on demand) services. This diagram also shows that usage records may be created for the selection and insertion of commercials and interaction with advertisements. This example shows that the money that is paid by the advertiser may go to the IPTV service provider.

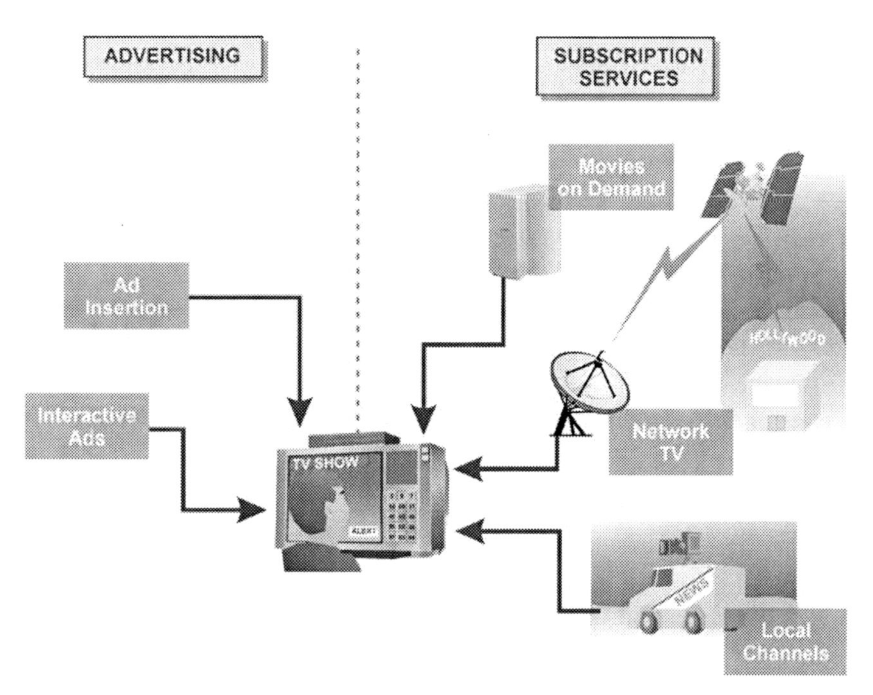

Figure 1.3, IPTV Television Services

Data Communication

Data communication is the transmission and reception of binary data and other discrete level signals that can be represented by a carrier signal that can represent the discrete level (usually on-off levels) for signal transmission. Data communication providers have networks and transmission lines that can cover relatively large geographic areas.

Data communication services include basic access transmission fees and information processing (storage and hosting). Transmission services may include transmission rates, guaranteed quality of service levels and data transmission quantity levels. Basic access fees may include a physical connection (such as a DSL line) and an Internet access fee. Some data transmission services may come with a guaranteed quality of service level that may include minimum data transmission rates and packet loss rates.

Data processing fees may include data storage, account hosting and other network configuration charges. Data storage fees may be in the form of Megabytes (MB) or Gigabytes (GB) stored. Hosting fees may include domain management (address mapping), maximum

Figure 1.4 shows typical data communication service providers typically bill for data transmission and data processing services. This diagram shows that a data subscriber is billed for basic data access service and additional levels of transmission service that may be provided including higher speed access, guaranteed quality of service (QoS) and amount of data transferred. This example shows that the data communication system routers and servers create event records that are combined to produce billing records.

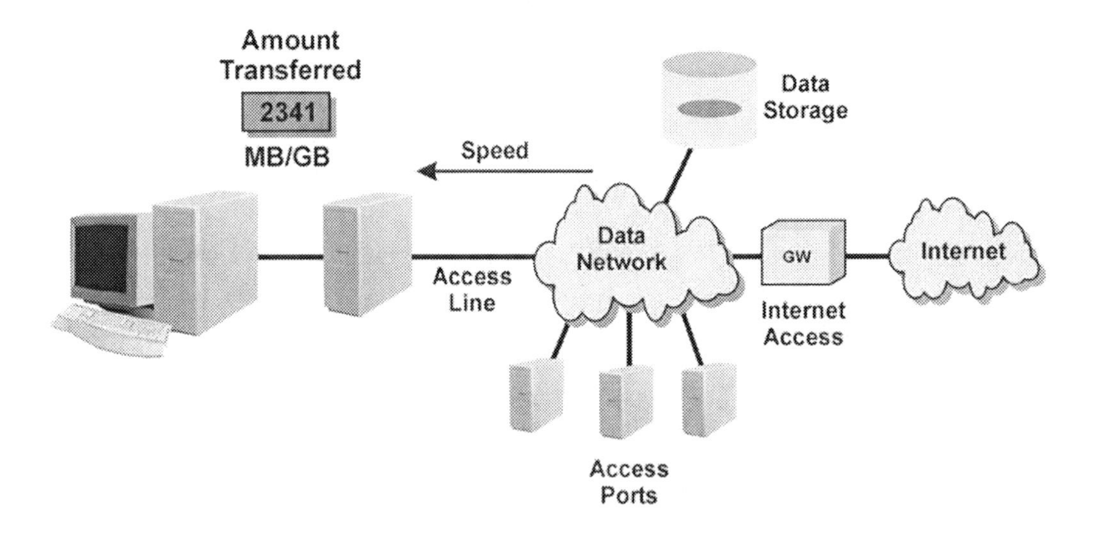

Figure 1.4, IPTV Data Communication Services

Telephony

Telephony is the use of electrical, optical, and/or radio signals to transmit sound to remote locations. Generally, the term telephony means interactive communications over a distance. Traditionally, telephony has related to the telecommunications infrastructure designed and built by private or government-operated telephone companies.

Telephone systems maintain connections between predetermined points in a communication network (telephone lines). Telephone systems interconnect through many systems and locations throughout the globe.

Telephony services may include local, long distance, international, toll free (freephone), prepaid, calling party pays (CPP) and other types of voice communication services. Customers typically pay for telephone service that is a combination of basic access fees and usage fees.

Figure 1.5 shows typical telephony services include access and network services. Basic access services include telephone access to local and long distance networks. Network services include toll free (freephone) access, prepaid, and calling party pays services that may be originated in other systems. This diagram shows that telephony communication services requires interconnection to a variety of public telephone communication providers and that a single usage record may be created from usage events from many different networks. To settle call charges through these networks, billing records are sent to a clearinghouse.

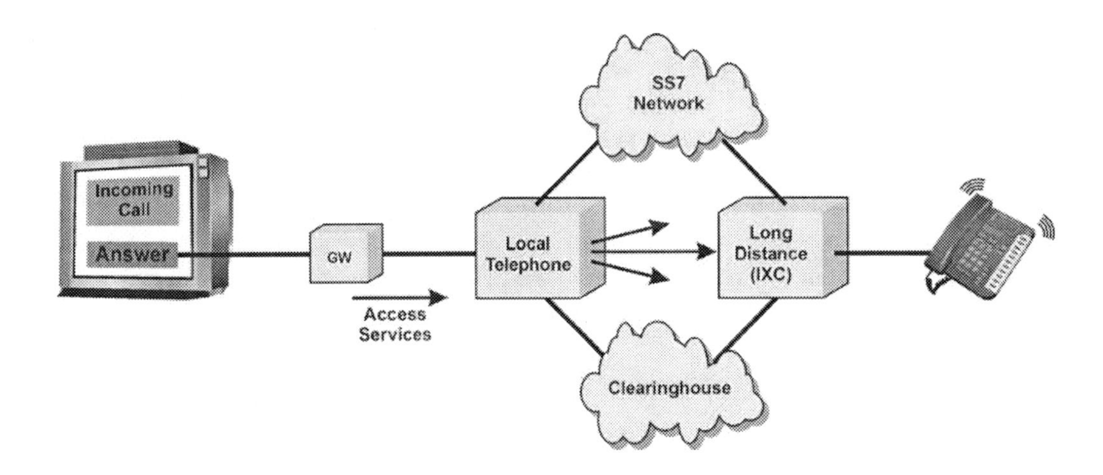

Figure 1.5, IPTV Telephony Services <ag_IPTV_Telephony_Services

Information Services

Information services involve the processing of information that is transferred through a communications system. Information services may be provided through a variety of systems including television, data communication and telephone networks. Information service providers add value to information by acquiring, storing, transforming, processing, retrieving, utilizing or making available information via communication systems.

Information services include gaming, news services, e-commerce, music downloads and other types of information processing services. Customers typically pay for information services through a combination of basic service access fees and charges for access to view or use media. The fees for information services may be directly billed to the customer (such as an e-commerce purchase) or it may be billed through an access network (such as a game that is downloaded through a service provider).

Information services or data manipulation that may be performed by the service provider or by other companies is called application service providers (ASPs). In return for allowing customers to access the information from other information service provider companies, some of the revenue generated from these information services is shared with the access provider.

Information services may be provided through variety of systems and access devices that have different display and navigation capabilities. This means that information service providers must adapt information into a variety of formats.

Information services are provided to users rather than specific devices. To identify users, account codes and security keys are assigned to uniquely identify the user and the services they request and receive. Service usage records are assigned to these account IDs.

Figure 1.6 shows how information services may be provided through a variety of access networks to different types of access devices. This diagram shows that information services are provided to users rather than devices so that an account code rather than a specific type of access device uniquely identifies each user. This example shows that the billing for information providers may be performed by direct billing from the information service provider or it may be billed and revenue shared by the access provider.

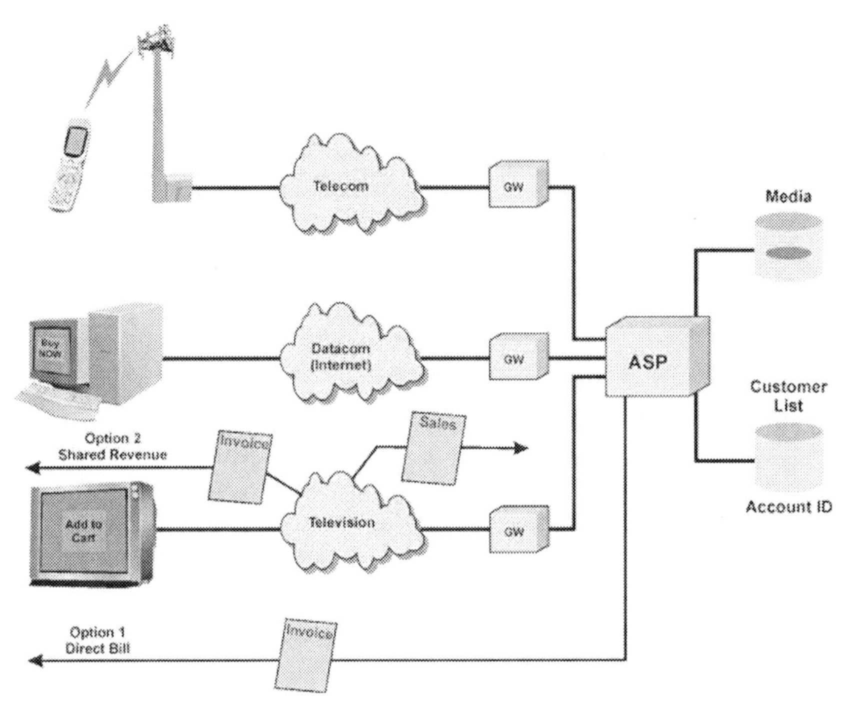

Figure 1.6, IPTV Information Services

Billing Systems

Billing systems are composed of interfaces (Network, Marketing, Customer Care, Finance, etc.), computers, software programs and databases of information. Computers are the hardware (computer servers) and operating systems are used to run the programs and process. Network interfaces are the hardware devices that gather accounting information (usage) from multiple networks, convert it into detailed billing records, and pass it on to the billing system. Billing systems databases hold customer information; usage call detail records, rate tables, and billing records that are ready to be invoiced.

The key functional parts of a billing system include creating usage records, event processing, bill calculation, customer care, payment processing, bill rendering and management reporting. In addition to the basic billing system functions, billing systems share information with many other business functions such as sales, marketing, customer care, finance and operations.

Billing charges are determined by events that occur in a communication system. Billing events can originate from many sources: a media gateway, a media server, a content aggregator or a visited partner's network and they must be converted into a standard format.

A typical billing process involves collecting usage information from network equipment (such as media servers, access devices and set top boxes), translating and formatting the usage information into records that a billing system can understand, transferring these records to the billing system, assigning charge fees to each event, creating invoices, receiving and recording payments from the customers.

Figure 1.7 shows an overview of a billing and customer care system that can be used for IPTV communication services. This diagram shows the key billing steps. First, the network records events that contain usage information. This example shows that billing information usually includes a user identification code, service request date and time, destination address, source address, media type, usage duration or amount of service that is provided. Next, these events are combined and reformatted into a single usage detail record (UDR). Because these events only contain network usage information, the identity of the user must be matched (guided) to the event detail record and the charging rate for the service must be determined. After the total charge for the service is calculated using the service rate for that particular user, the billing record is updated and is sent to a bill pool (list of ready-to-bill usage records). Periodically, a bill is produced for the customer and as payments are received, they are recorded (posted) to the customer's account. The charge is then "journalized"; i.e. it is assigned a financial account.

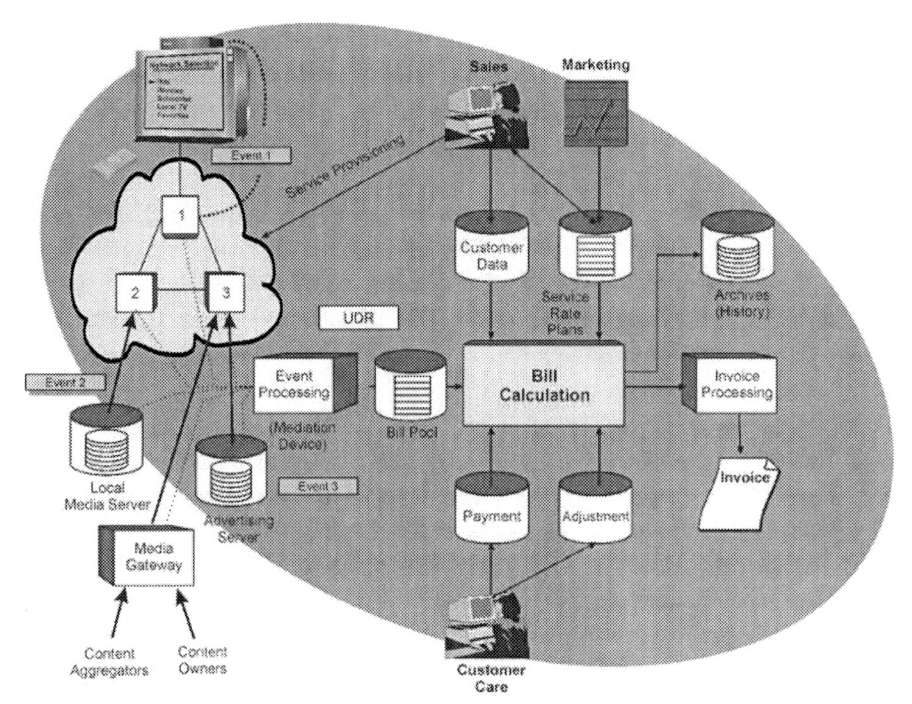

Figure 1.7, Billing and Customer Care System

Event Sources and Tracking

Events with regard to a billing system are measures of network or media usage. Events can be stored in a storage device (data collector) for transfer at predetermined time intervals, when a specific value has been reached (event trigger) or when the billing system requests the information (called polling).

Some common IPTV event sources include: access requests to media gateways, multicast routers, application servers or the use of the media content itself. Media gateways are devices that convert media from one format (such as satellite digital video) to another format (such as IP video). Media gateways can identify the time a service is requested and ended. Multicast routers are intelligent switches that can copy and forward packets to multiple destinations based on requests to join multicast groups. Multicast routers can track the time a user has requested to join a group (such as to watch a television program) and the amount of data that is routed between two ports over a period of time. Application servers are computers that process information at the request of a customer (called a client). Application servers can track the beginning (launching) and termination of an application. An example of an application may be the user's selection of an item or a link on an interactive advertising message. The usage of media files themselves may trigger the sending of event information. For example, if a registered stored video file is played, this may trigger an event message that is sent to a predetermined server so that the usage of the media file can be recorded.

It is also possible for IPTV systems to have multiple accounts that operated on the same access device. While the access device (such as a set top box) may have a unique physical address (MAC address), each user must be identified by other unique information such as a removable identification card (e.g. SIM card) or login identification and password codes.

Figure 1.8 shows how it is possible for multiple accounts to be serviced by the same access device (such as an IPTV set top box in this example). This example shows that a television is shared by three roommates who each have their own IPTV account. Each user uses an access code each time they want to obtain access to premium content. The billing system is able to differentiate between users based on the login access code (account ID) rather than the unique network address of the physical device.

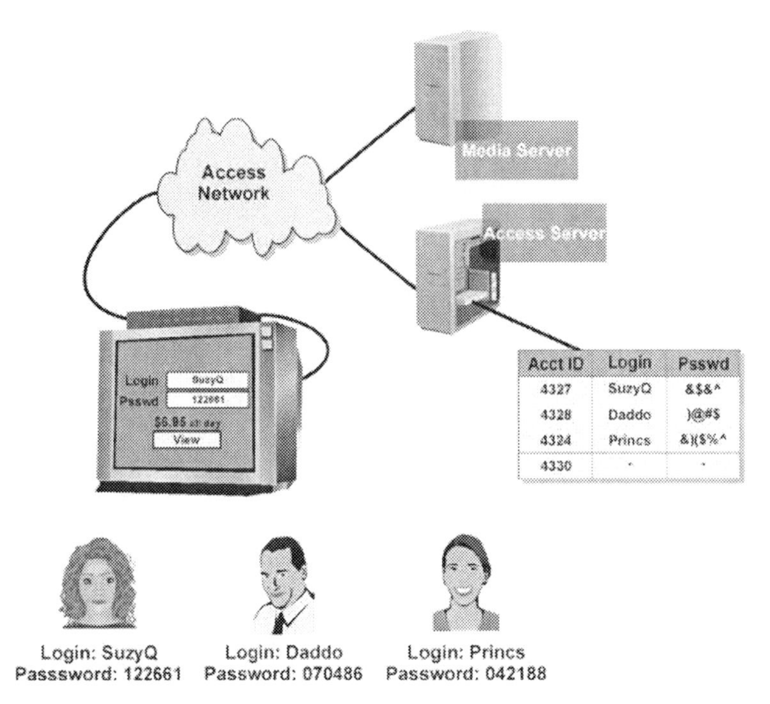

Figure 1.8, Multiple IPTV Accounts for the Same STB Access Device

Mediation Devices

A mediation device is software or equipment that filters information and insulates the enterprise applications from the network elements. As such, the mediation device receives, processes, and reformats usage information in a communications network to a suitable format for processing by one or more billing and customer care systems. This processed information is either continuously or periodically sent to the billing system. Mediation devices are commonly used for billing and customer care systems as these devices can take non-standard proprietary information from switches and other network equipment and reformat them into messages billing systems can understand.

Switches usually record usage information (e.g. service connection time) in a format that is often proprietary to the manufacturer of the network device. Each record may be variable lengths and several events (e.g. media types) may be recorded in the same system for a single service.

There are other network parts or devices that may be involved with providing a service or providing value added services (VAS). These devices also produce event detail records yet in different formats.

Usage Detail Records (UDRs)

Billing information regarding specific usage events resides in usage detail records (UDRs). UDRs hold a user identification code (who), origination address of the device (this may indicate the user), time of day the usage was provided (when), the usage type and its details (what), the duration or quantity of the service (how much), the connection location(s) of the service (where), and the cause of event recording (why). UDRs hold billing record information about traditional services (such as television viewing) along with non-traditional services such as information services that are provided by other companies.

Figure 1.9 shows the general process that is used to identify and rate (bill) a service. This diagram shows that a usage detail record evolves as it passes through the rating process. In the first step, all the events for a specific service are combined and adapted to a common format. The usage detail record is then guided to a specific account. Using the account identification code, a rate plan is discovered the unit (usage) and/or fixed amounts (per event) charging rates are gathered and calculated. The new information (rate plan, usage charge amount) is added to the usage detail record and it is moved to the bill pool, as it is ready to be billed.

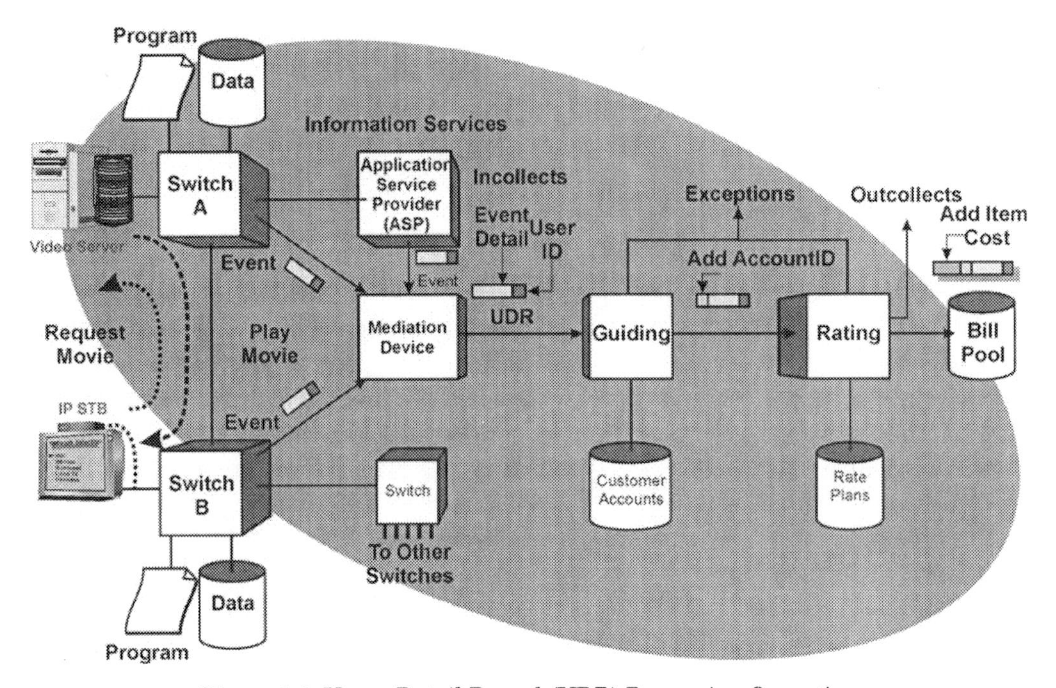

Figure 1.9, Usage Detail Record (UDR) Processing Operation

Incollects

Incollects are billing records that are received by service provider A from service provider B for services provided by B to A's customers. An example of an incollect is a billing record for the viewing of a movie that was provided via another network.

Outcollects

Outcollects are usage billing records that a network operator sends to other companies for the services they provided to customers that are not registered in the local network (such as providing access to movies to users registered in other systems). An example of an outcollect is a billing record for viewing a movie by an unregistered individual.

Rating

Rating is a function within the billing system that assigns a rate (cost parameter) to a usage record. Rating is performed by a rating engine (a software program).

Rating typically involves using the originating number or network address, terminating number or destination network address, date the service was used, amount of usage or time period, usage type and tax jurisdiction to determine who the customer is and the initial charge assigned to the usage record. The actual cost of the usage record may be adjusted based on volume discounts or other rate plan considerations.

After a UDR has been rated and the actual charge for the service is calculated, the call detail record is moved into a group of billing records that are ready to be invoiced called a "bill pool".

Figure 1.10 shows how a rating engine of a billing system can calculate the fees associated with specific usage events. This diagram shows that the rating engine receives usage records from the network element, incollects from other companies and other billing records (such as adjustments). The rating engine first identifies the account associated with the usage records and checks for duplicate records. The rate table for the customer account is selected for these records and the appropriate fee is calculated for each record. After the calculated rate is added to the usage detail record, it is sent to the bill pool to await the aggregation and processing into an invoice.

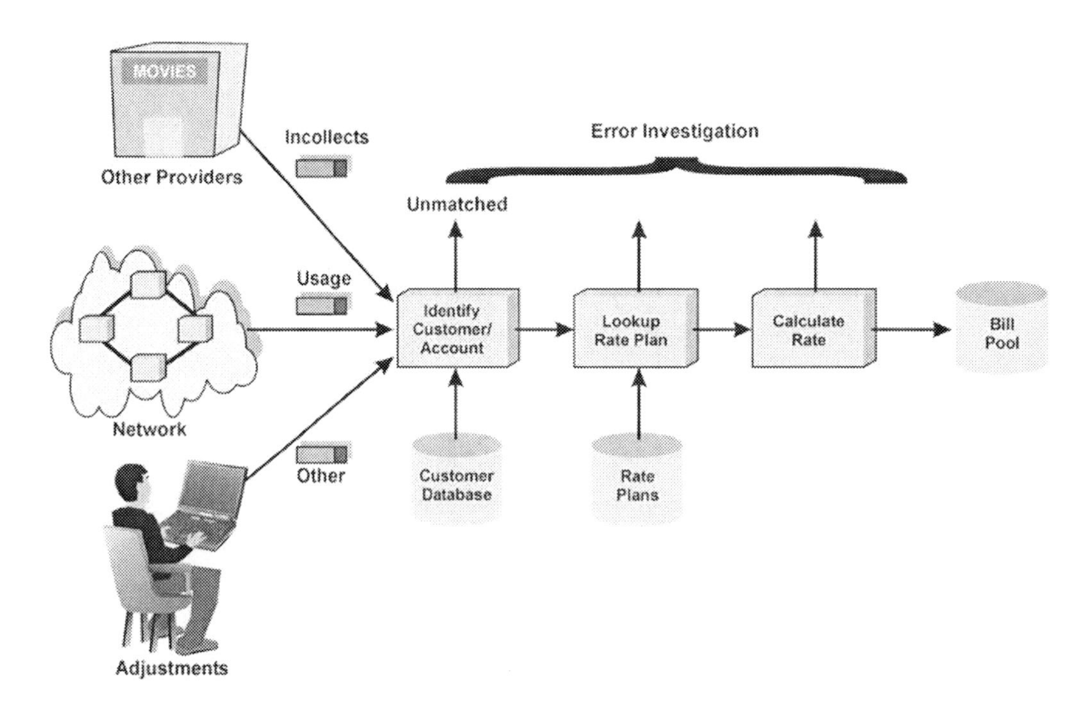

Figure 1.10, IPTV Billing Record Rating

It may be necessary to divide the UDR into several component parts. For example, selecting an international television program may be divided into a movie access fee and international program surcharge (to pay the international provider).

Guiding Process

Guiding is a process of matching a UDR to a specific customer account. Guiding uses the UDR identification information such as the calling telephone number or user identification login code to match to a specific customer account.

Rate Plans

A rate plan is the structure of service fees that a user will pay to use services. Rate plans are typically divided into monthly fees and usage fees.

Rate plans may include volume discount or threshold levels (tiers) of service that change the usage rate based on factors such as total quantity used, number of uses or interaction with specific types of services (such as viewing local video content).

Invoicing

Invoicing is the process of gathering (aggregating) and adding up all of the billing records associated with a specific account during a billing cycle in bill pool, applying recurring charges (e.g. monthly charges) and totaling all the charges.

The invoicing process typically starts by gathering all of the billing records for a specific customer from the bill pool that occurred during a specific time period (the billing cycle). These charges are converted into a form that can be displayed on the invoice (such as movies viewed or total minutes used). As the records are gathered from the bill pool, they are either transferred, deleted or marked as invoiced so these billing records are not used again.

After all the billing records during the billing cycle have been gathered, additional charges and credits such as monthly fees, taxes and billing adjustments are added to the invoice details. All the charge details are then totaled to complete the invoice.

Figure 1.11 shows the basic invoicing process involves creating an invoice record and aggregating usage records from the bill pool into a billing detail or summary section. As usage records are gathered from the bill pool, they are removed or marked to keep them from being transferred again. Additional non-usage fees (e.g. monthly service fees) and taxes are added to the invoice record. The information is then totaled to provide the customer with total amount due for this invoice.

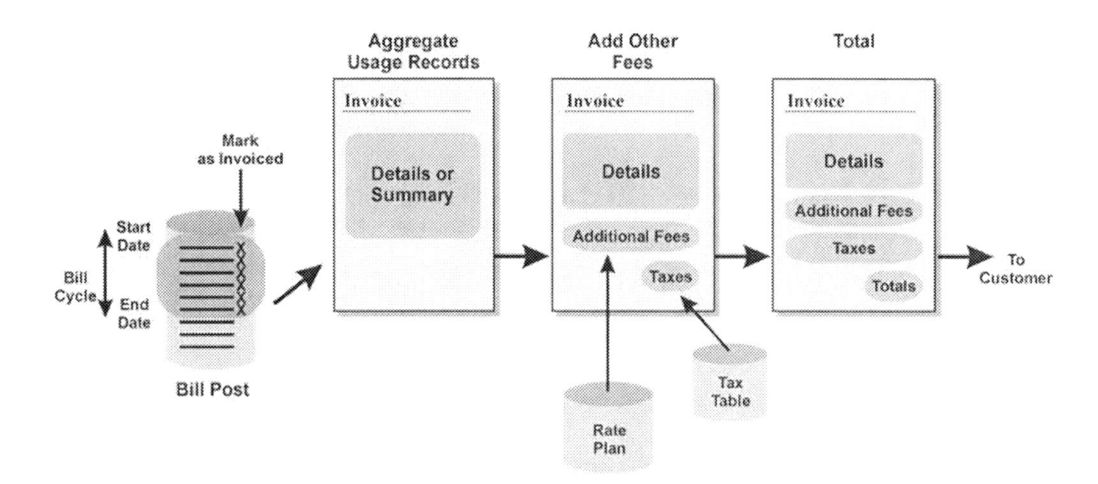

Figure 1.11, Invoice Processing

Rendering

Bill rendering is the conversion of billing information into a form that a human can view, hear or sense. Some of the options for bill rendering include bill printing, sending invoices via email, electronic data interchange (EDI), stored media, providing billing information via audio announcements by an interactive voice response system and web billing. An example of bill rendering is the conversion of a billing invoice into an image that can be displayed on a computer monitor.

Bill printing may be performed by a service bureau that specializes in printing and mailing invoices. Invoices may be converted into an electronic format that may be sent via email. Invoice emails may be in text only or graphics (HTML) formats. Invoices may also be formatted into an electronic data interchange (EDI) format that allows for the direct transfer to customer's computers. A common EDI format is extensible markup language (XML) and the use of EDI files may allow easy entry of invoices into the customer's accounting system. It is sometimes requested that detailed billing records be provided on stored media such as magnetic tapes and CD ROMs.

Billing systems may use interactive voice response (IVR) systems to allow customers to call in and receive audio updates about their billing status. The use of IVR systems can dramatically reduce the amount of calls handled by customer service representatives (CSRs) as many of the calls are billing related questions. Billing systems may be integrated with web servers to allow for web based billing. Web billing is the integration of an accounting system with Internet web systems. Web billing typically allows customers to view and possibly pay invoices through a web site. Web billing may have real time or near-real time (delayed) posting of customer transactions with the accounting system.

Figure 1.12 shows some of the different options available for bill rendering and distribution. This example shows that bills can be printed and mailed, created in text form and emailed, converted to stored media such as CDROM or tape, can be formatted for web viewing or converted to a standard EDI format to be directly sent to a customer's computer server.

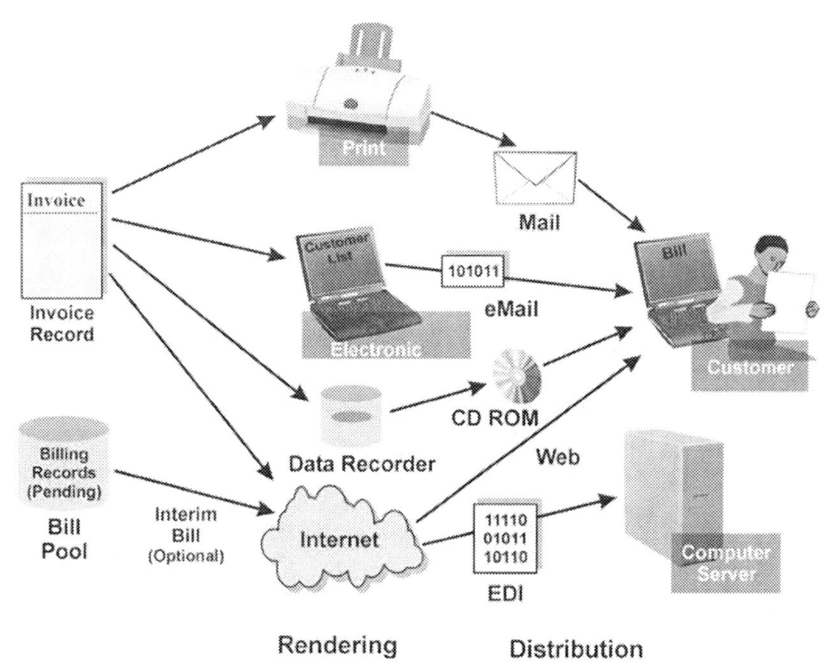

Figure 1.12, Bill Rendering Options

Customer Care

Customer care is the processes and communication that occurs between customers and companies to enable customers to resolve problems and successfully obtain products and services from the company.

Customer relationship management is the process or system that coordinates information that is sent and received between companies and customers. CRM systems are used to schedule activities, allocate resources, and help control the sales activities within a company.

Payment Processing

Payment processing is the tasks and functions that are used to collect payments from the buyer of products and services. Payment systems may involve the use of money instruments, credit memos, coupons, or other form of compensation used to pay for one or more order invoices. Payment options include cash, check, credit card, paypal, credit memos, coupons and electronic funds transfer (EFT).

Payment processing can be complicated as the receipt of funds is typically applied to specific invoices or on account and the payment method may not include the invoice or even the account number.

Cash and checks may be received at company locations or at authorized agents. Credit card payments are processed through merchant processors who charge a fee (typically 1.5% to 4%) for the processing of transactions. Customers may apply credits (credit memos) or coupon codes to be applied as a portion of their payment. Customers may also use electronic funds transfer (EFT) to directly transfer funds.

Figure 1.13 shows some of the different options available for bill payment. This example shows that customers can make cash payments to a company office of authorized agent, can send checks to the company, can pay via credit cards, via a 3rd party financial processor (e.g. paypal) or through electronic funds transfer (transfer fund). This example shows that payments on account are applied to specific invoices and when the invoices are paid in full, they are marked as paid and are not available for additional payments to be applied.

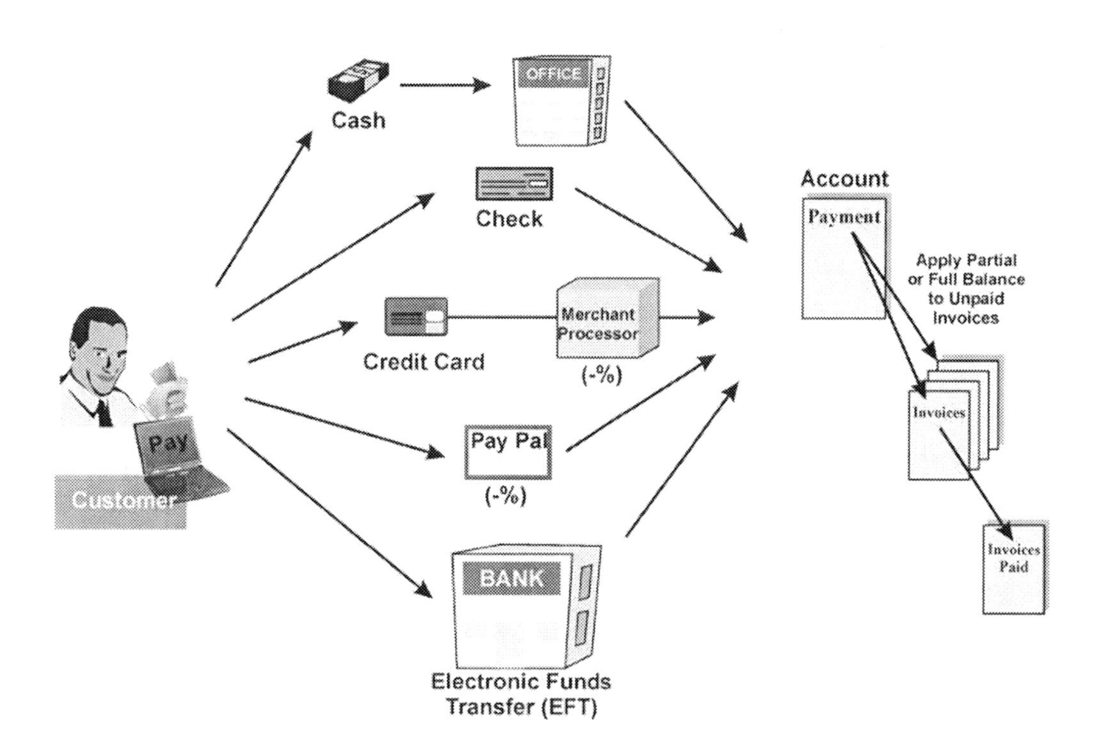

Figure 1.13, Bill Payment Options

Billing Reporting

Billing reporting systems provide information about specific events (such as call or service usage) or summary information that allows management to make decisions about products and services such as product management, fraud management and revenue assurance.

Product Management

Product management is the process of assigning and tracking specific tasks and functions related to ensuring the success of products or services.

Fraud Management

Fraud management is the set of processes and steps taken to identify, minimize and correct the unauthorized use of products or services.

Revenue Assurance

Revenue assurance is the process of reviewing processes, systems and data that are associated with revenue streams to ensure billable services are correctly recorded and collected. A revenue stream is a source of money or value received that is associated with the sale or providing of services or products or can be associated with a specific type of sales or marketing process.

Billing Service Bureau

A billing service bureau is a company that provides billing services to other companies. The services provided can range from billing consolidation to complete billing operations that include gathering billing records, processing invoices, mailing or issuing the invoices, and posting payments.

Billing Standards

Billing standards define the measurements, record format and the methods of transfer for billing related information within a network. Because of deregulation of the telecommunications industries around the world, new services are being offered by network operation. As a result, billing standards are continually being revised and they are converging. Because companies can use different billing standards or different revisions of billing standards, clearinghouses often provide translation services between different billing standard formats.

The use of industry standard billing systems is important to processes such as revenue sharing, partner settlement, authorizations and denials of service, as well as the ability to adequately use the data collected for marketing and customization of services. Configuring systems to accept standardized input will allow an easier operations model for not only service providers, but also trading partners that are involved in the value chain of the service. Settlement among various partners that is of greater complexity of voice or data service will drive the need to have industry collaboration on an accounting standard that will allow a reduced time in the market entrance of new service and features. Not just getting data from the network of being able to identify the service associated with the usage and handle the usage accordingly.

There are many billing standards that have been developed for telecommunications networks. Because the services offered by different types of network operators (e.g. cable television compared to local telephone companies) are beginning to overlap, billing standards are also converging.

Some of the billing standards that may be used for or linked to IPTV billing systems include Internet Protocol Detail Record (IPDR), automated message accounting (AMA), Cellular Intercarrier Billing Exchange Roamer (CIBER), and Transferred Account Procedures (TAP).

Internet Protocol Detail Record (IPDR)

Internet protocol detail record (IPDR) is a standard billing format that is used for exchanging information about and over IP data communication systems. The IPDR billing standard allows for the use of variable length data packets and includes user definable fields to allow IPDR billing records to be adapted for many different types of services.

IPDR data records contain information related to an IP-based communication session. This information usually contains identification information of the users of the service, types of services used, quantity measurement unit type (e.g. kilobytes or time), quantities of services used, Quality of Service parameters, and the date/time (usually relative to GMT) the services were used.

Exchange Message Record (EMR)

Exchange message record (EMR) is a standard format for the exchange of messages between telecommunications systems. The EMR format is often used for billing records. The records may be exchanged by magnetic tape or by other medium such as electronic transfer or CD ROM.

Automatic Message Accounting (AMA)

Automatic message accounting (AMA) is a standard record gathering and billing format that is used primarily by local telephone companies to process billing records and exchange records between systems. The AMA format was created by BellCore and is now managed by Telcordia.

Carrier Interexchange Billing Exchange Record (CIBER)

Carrier interexchange billing exchange record (CIBER) is a billing standard designed to promote intercarrier roaming between cellular and other wireless telephone systems. The CIBER format is developed and maintained by CiberNet. The Cellular Telecommunications Industry Association (CTIA) owns Cibernet.

Figure 1.14 shows some of the information (fields) contained in a type 22 CIBER record. This example shows that the type 22 Ciber record field structure has been updated from the previous type 20 record structure to include additional fields that allow for telephone number portability (enabling telephone number transfer between carriers). This list shows that fields in the Ciber record primarily include identification of airtime charges, taxes, and interconnection (toll) charges.

Type 22 Record - Sample of Fields

- Home Carrier SID/BID
- MIN/IMSI
- MSISDN/MDN
- ESN/IMEI
- Serving Carrier SID/BID
- Total Charges and Taxes
- Total State/Province Tax
- Total Local Tax
- Call Date
- Call Direction
- Call Completion Indicator
- Call Termination Indicator
- Caller ID
- Called Number
- LRN
- TLDN
- Time Zone Indicator
- Air Connect Time
- Air Chargeable Time
- Air Rate Period
- Toll Connect Time
- Toll Chargeable Time
- Toll Carrier ID
- Toll Rate Class

Figure 1.14, CIBER Billing Record Structure

Transferred Accounting Process (TAP)

Transferred accounting process (TAP) is a standard billing format that is primarily used for 3rd generation WCDMA wireless, global system for mobile (GSM) cellular, and personal communications systems (PCS). As of 2003, the versions of TAP TAP 2, TAP 2+, NAIG TAP 2 and TAP 3. Each successive version of TAP provided for enhanced features.

Due to the global nature of 3G wireless and GSM, the TAP billing standard provides solutions for multi-lingual and multiple exchange rate issues. TAP3 was released in 2000 as a significant revision of TAP2. TAP3 has changed from the fixed record size used in TAP2 to variable record size and TAP3 offers billing information for many new types of services such as billing for short messaging and other information services. The GSM association at www.GSMmobile.com manages the TAP standard.

Network Data Management – Usage (NDM-U)

The network data management – usage (NDM-U) is a standard messaging format that allows the recording of usage in a communication network, primarily in Internet networks. The NMD-U defines an Internet Protocol detail record (IPDR) as the standard measurement record.

Because Internet services are now offered in almost all communications systems, the IPDR record structure is very flexible and new billing attributes (fields) are being added. The NMD-U standard is managed by the IPDR organization at www.IPDR.org.

Data Message Handler (DMH) Interim Standard 124 (IS-124)

The data message handler (DMH) interim standard 124 (IS-124) allows for the real time transmission of billing records between different systems, primarily between wireless systems in the Americas. IS-124 messaging is independent of underlying technology and can be sent on X.25 or SS7 signaling links.

The development of the standard is primarily led by Cibernet, a division of the cellular telecommunications industry association (CTIA).

Extensible Markup Language (XML)

Extensible markup language (XML) is a software standard that is used to define exchangeable elements of a file such as a web (HTML) page. Extensible Markup Language was developed in 1996 by the World Wide Web Consortium (W3C). It is a widely supported open technology (i.e. non-proprietary technology) for data exchange. XML documents contain only data, and applications can display data in various ways. XML permits document authors to create their own markup for virtually any type of information. Therefore, authors can use XML to create entirely new markup languages to describe specific types of data, including mathematical formulas, chemical formulas, music and recipes.

Extensible Rights Markup Language (XrML)

Extensible rights management language is a XML that is used to define rights elements of digital media and services. Extensible Rights Markup Language was initially developed by Content Guard and its use has been endorsed by several companies including Microsoft. The XrML language provides a universal language and process for defining and controlling the rights associated with many types of content and services.

Because XrML is based on extensible markup language (XML), XrML files can be customized for specific applications such as to describe books (ONIX) or web based media (RDF). For more information on XrML see www.XrML.org.

Service Usage Fees

Service usage is the measurement of a type of usage of content or service. IPTV services can be a combination of television/video, data communication, telecommunication, and information services so there are thousands of potential service identification codes in a single IPTV billing system. Service usage fees may be a mixture of recurring (periodic) or one-time (single) charges that occur over a billing cycle.

Periodic Charges

Periodic charges are fees that are associated with a product or service that is assessed on a regular interval (i.e. monthly, quarterly, annually).

Periodic charges may be prorated based on the duration of service used during the billing cycle. Prorating is the process of fractionalizing charges for a partial period. In order to determine the partial charge, the number of days that the service was available to the customer is multiplied by a per-day-charge called a "Multiplier".

Activation Fees

Activation fees are one-time fees that are charged for the initial setup of communication service. Activation fees are also called "setup fees."

Equipment Leasing Charges

Equipment leasing are the charges assessed for the use of equipment. Equipment leasing options for IPTV include set top boxes, modems, routers, televisions and any other type of equipment that can be used in the customer's premises to allow for the reception, distribution and display of IPTV media.

Penalties

Penalties are charges that are assessed for actions or service usages that fall outside the agreed limits of service usage. Penalties can include early contract termination fees, charges for lost equipment or the usage of services by unapproved devices (e.g. media storage devices).

Late Fees

Late fees are charges that are added to an account or invoice for failure to pay charges by the due date or series of due dates that have been previously agreed upon by the customer.

The assessment of late fees may be regulated so that pyramiding of late fees does not occur. Pyramiding late fees is the charging of late fees from a previous invoice period so that the payments applied are not enough to cover the current late fees resulting in the assessment of additional late fees.

Disconnection Charges

Disconnection charges are fees that are assessed for the disconnection of a service. Disconnection fees may be charged if the user fails to pay the account balance and the service is disconnected for non-payment; or if the customer has requested to disconnect service before the term of the service agreement is complete.

Re-Activation Charges

Reconnection charges are fees that are assessed for re-connecting a service that was previously disconnected. Reconnection charges may be lower than new service activation charges because much of the setup work may have already been done and customer information is already available which reduces the amount of effort and resources that are required to reactivate the service.

Discounts

Discounts are reductions of pre-established fees or tariffs that are given for specific reasons. Discounts may be in the form of a specific amount or they may be based on a percentage of an item's price or invoice amount. Discount types may be coded using specific identification codes and discount rates may be applied based on the specific type of sale or customer category using a discount schedule. The account representative may grant discounts on an ad-hoc basis to a specific customer.

A discount schedule is an itemized list or table that provides pricing discount information for products or services. The price discount schedule will usually include the amount of discount based on usage volumes, quantity of product purchased, and the types of customers that qualify to receive the discounts (e.g. wholesale, education, retail).

Item Usage Fees

Item usage fees are the charges or assessments for the authorization to access (the service may not actually be used) or usage of products and services. Item usage fees can be based on duration, volume or type of service.

Duration Based Usage Charges

Duration based usage charges are fees that are assessed for the amount of time a service is authorized for use. An example of a duration based usage charge is thc providing of access (or the right to access) to specific items or services for a defined time usage criteria (e.g. authorization to watch a movie during a 24 hour period).

Volume Based Usage Charges

Volume based usage charges are fees that are assessed for the amount of service that is consumed. An example of a volume based usage charge is the providing of quantity of service (or the ability to consume a specific volume) to specific items or services for a defined quantity (e.g. the amount of data transferred, or number of movies downloaded).

Quality of Service (QoS)

Quality of service (QoS) fees are charges or assessments for providing services at a specified level of performance. QoS fees may vary based on the requested quality of service level and the actual delivered quality of service level. For example, if the customer requests a communication service guaranteed data transmission rate such as a committed bit rate (CBR) of 1 Mbps, this may be an upgrade to an available bit rate (ABR) service data connection.

Overage Charges

Overage charges are fees that are imposed or added to an account for having exceeded the allocated amount of usage for a defined level, or quantity of service, or time period.

Data Storage Fees

Data storage fees are charges for the allocation and/or usage of storage media. Storage fees may be charged for raw data storage or for the storage of particular types of media (such as movies, pictures and voice mailboxes). Data storage fee rates may vary based on the reliability (such as backups) and data transfer performance available from the data storage device.

Figure 1.15 shows some of the common types of service usage metering. This diagram shows that the types of usage metrics may include the amount of time a service has been used, how much of a service may be used, the number of times a service has been used or activated, the type of use (e.g. single viewer or public viewers), quality of service (e.g. high resolution or low resolution) or the location of the service access point (e.g. home or at a visited/away location).

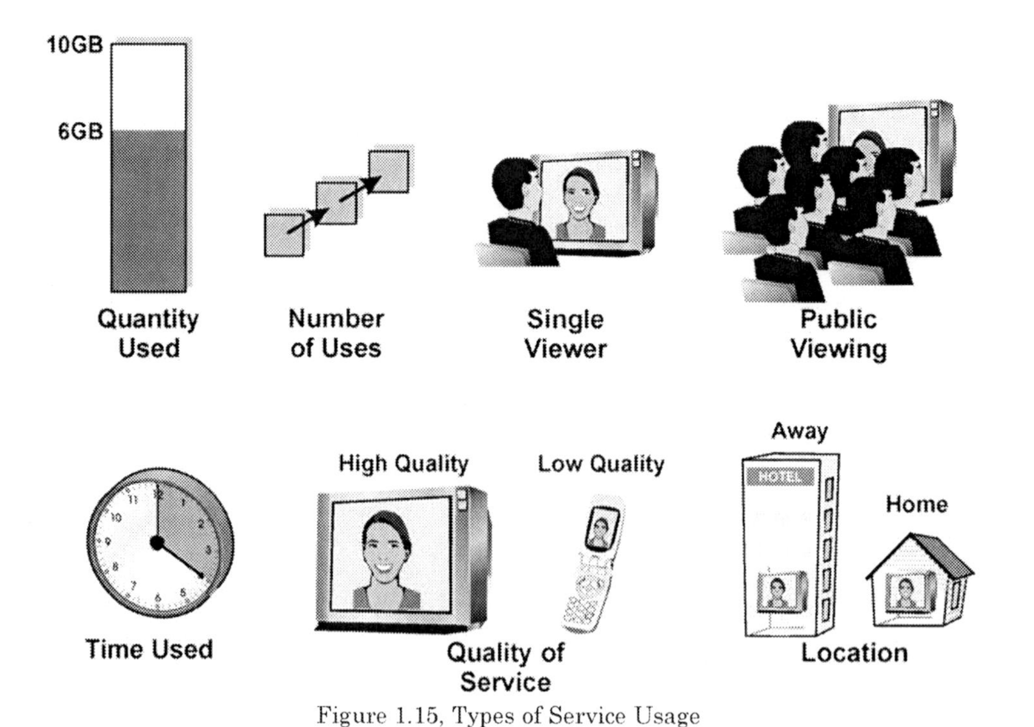

Figure 1.15, Types of Service Usage

Advertising Fees

Advertising is the communication of a message or media content to one or more potential customers. In addition to revenue from providing services or content to subscribers, additional revenue may be received from advertisers.

Ad insertion is the process of inserting an advertising message into a media stream such as a television program. For broadcasting systems, advertising messages are typically inserted on a national or geographic basis that is determined by the distribution network. For IP television systems, Ad inserts can be directed to specific users based on the viewer's profile.

IPTV service offers the option to insert advertising messages into programming at a variety of distribution points. Advertising services for IPTV systems include network advertising, (to all viewers), regional advertising (spot ads), ads to specific viewers (addressable ads) and interactive ads (ads that the viewer can immediately react or respond to).

Charging an advertiser for the delivery of advertising messages includes fees for ad impressions and ad selections. Ads may be inserted on a guaranteed or space available basis.

Ad Impressions

An ad impression is the presentation of an advertising message or image to a media viewer. Charging for ad impressions on an IPTV system is different because it is possible to pay based on the actual impressions to specific viewers. The better the profile of the viewer matches the desired audience of the advertiser, the more valuable the ad impression becomes.

Ad Selections

Ad selection is a choice or option selected (e.g. clicked) by a viewer during an ad impression. Unlike broadcast television services, IPTV systems typically allow for a return communication channel along with the ability to redirect programming to specific advertising segments. IPTV service providers can charge for ad links selected by customers and because ad selections indicate an interested prospective customer, ad selection value can be hundreds of times higher than the value of ad impressions.

Equipment Purchase Charges

Equipment purchases are charges for devices that a customer may be allowed to control and keep. Equipment purchases may be required or optional.

Equipment Insurance Fees

Equipment insurance is charges for the protection or financial recovery of assets that may be damaged or lost for specific reasons.

Adjustments (Credit or Debit)

Adjustments are changes to item charges, invoices or account balances. Adjustments are typically made to correct errors or to provide satisfaction to customers for billing errors or services that did not meet expectations. Some of the common adjustments include charge waivers and refunds.

Because accounting reports are periodically created for management and tax purposes, adjustments are made rather than changing previous invoice information as these changes would alter reports that were previously produced. Adjustments therefore appear on the next bill cycle's invoice.

Charge Waivers

Charge waivers are credits that are applied to an account or invoice to reverse a charged item or fee. Waivers may be full or partial and appear on the next bill cycle's invoice.

Refunds

Refunds are credits that are applied to an account or invoice to reverse a previously received payment. Refunds are actual payments made directly to the customer.

Finance and Interest Charges

Finance and Interest charges are the fees charged for overdue balances.

Bill Rendering (Paper or CD ROM) Fees

Bill rendering fees are charges for the production of billing records in specific formats such as paper or other forms of stored media (e.g. CD ROM). The fees for bill production (bill rendering) may vary based on the detail level that is included in the bill format.

Network Access Charges

Network access charges are fees or assessments for the usage of particular network resources such as communication circuits, channels or access points.

Circuit Charges

Circuit charges are fees or assessments for the allocation or dedication of communication paths between access points in a communication network.

VPN and PVC Charges

Virtual Private Network (VPN) and Permanent Virtual Circuit (PVC) charges are fees or assessments for the setup and management of logical (virtual) connections in a communication system.

Port Charges

Port charges are fees or assessments for the providing of access points (ports) into a communication network.

Taxes

Taxes are charges or levies assessed by a government agency or authority for services that are provided or products that are sold and sales that are defined by the government as a taxable commodity. There can be many types of taxes imposed on a service provider and the calculated tax fees depend on the type of service provided, the location of the service and potentially other criteria. Taxes may be assessed by a combination of national, regional and local authorities.

Regulatory Surcharges

In addition to taxes, communication regulatory agencies may impose additional fees (surcharges) to assist in the costs of management and development of public communication services.

Interconnection Access Surcharge

Interconnection access surcharge are fees that are collected to help develop access systems that allow customers to select alternative communication access providers.

Emergency Services Surcharge

Emergency services surcharge is the fee that is collected from a user of communication services to help operate emergency services such as Public Safety Answering Points(PSAP). PSAPs are facilities that receive and process emergency calls. The PSAP typically receives the calling number identification information that can be used to determine the location of the caller. The PSAP operator will then initiate and/or route calls to assist with the emergency situation.

Universal Service Charge (North America)

Universal service charges are fees that are collected from service users in a communication system to subsidize the construction of communication systems in rural areas to allow people in all areas within a country to have affordable access to communication services.

Security Deposits

Security deposits are asset collections that are controlled by a service provider or company for the assurance that a person or company will fulfill their obligations such as performing an action or payment for services.

Other Charges & Credits (OC&C)

Other charges and credits are fees or assessments for services or products that cannot be classified in other categories (e.g. miscellaneous charges).

Content Usage Fees

Content usage fees are the charges that are assessed for the availability or transfer of content. Content usage fees may be in the form of monthly licensing charges or per access fees from content producers, content aggregators or original programming royalties. The cost of content usage fees can exceed 50% of the collected program revenues.

Content Producers

Content producers are companies or developers of media content. Content producers may directly provide distributors or network providers with access to content.

Content Aggregators

A content aggregator obtains the rights from multiple content providers to resell and distribute through other communication channels. A content aggregator typically receives and reformats media content, stores or forwards the media content, controls and/or encodes the media for security purposes, accounts for the delivery of media and distributes the media to the systems that sell and provide the media to customers.

Original Programming

Original programming is content that is owned, developed and controlled by network operator who provides the media to its viewers. Original programming may be in the form of news, documentaries, education and other programming that is created for the network. The creation of original programming may reduce or eliminate content costs. However, original programming may still involve the payment of fees for the use of brands, actors or other images in the original programming content.

Shared Revenue

Shared revenue is a process where a company shares revenue for the providing of services or content. Shared revenue services allow a service provider to provide a portal (access point) and billing for the delivery of media content or information services and to split (share) the revenue with the content provider.

Usage Royalties

Usage royalties are fees that are paid to an author or composer for the right to use each copy of a work that is sold, performed, or produced under license of a exclusive right (such as patent rights).

License Fees

License fees are an amount charged or assigned to an account for the authorization to use a product, service, or asset. License fees can be a fixed fee, percentage of sales, or a combination of the two.

Integration with Legacy Systems

IPTV systems typically integrate multiple types of systems including television, telephone and data communication systems. As these systems provide services, billing records are created and they may be combined by a convergent billing system to provide customers with a single bill for multiple types of communication services.

Systems Integration

Systems integration is the process of defining, selecting, combining and configuring multiple types of systems to operate together to perform specific functions and/or services. Systems integration can range from porting (simple one-way connections) of systems to each other to full integration (two-way interactive processing) operation.

System integration commonly involves the connection of new systems with existing legacy systems. This integration allows the legacy systems to continue to operate while the new systems are installed, tested and validated.

Figure 1.16 shows how IPTV systems may combine telephone systems, data networks and data communication systems into a common IPTV system. This diagram shows that an integrated access device (IAD) may include connections for telephones, data networks and television signals. The IAD is setup to communicate with several different legacy systems through a shared broadband connection. To allow the IAD to interoperate with these different types of systems, gateways are used. The service provider will manage these multiple systems through an integrated network management system (NMS).

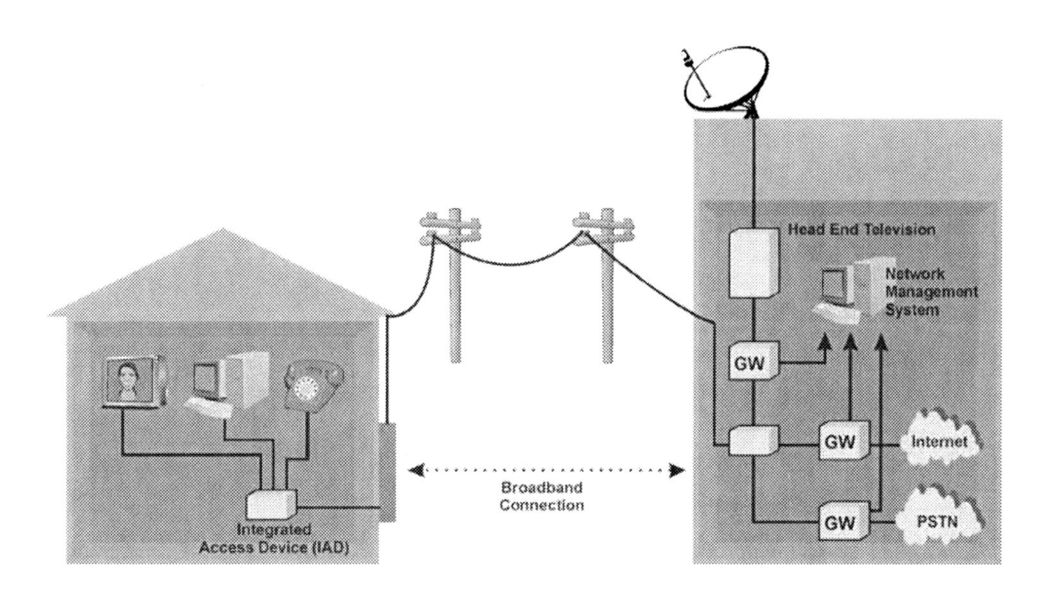

Figure 1.16, IPTV System Integration

Convergent Billing

Convergent billing is the combining of billing information for multiple types of services such as television, telephone service and data communication services. Convergent billing or systems may be tightly integrated or loosely tied together (sometimes called "stapled") to allow a customer to have a single billing and customer care access point.

Index

Althos Publishing Book List

Winter 2005-06

Product ID	Title	# Pages	ISBN	Price	Copyright
Billing					
BK7727874	Introduction to Telecom Billing	48	0974278742	$11.99	2004
BK7769438	Introduction to Wireless Billing	44	097469438X	$14.99	2004
Business					
BK7781359	How to Get Private Business Loans	56	1932813594	$14.99	2005
BK7781368	Career Coach	92	1932813683	$14.99	2006
Datacom					
BK7727873	Introduction to Data Networks	48	0974278734	$11.99	2003
IP Telephony					
BK7727877	Introduction to IP Telephony	80	0974278777	$12.99	2003
BK7781361	Tehrani's IP Telephony Dictionary, 2nd Edition	628	1932813616	$39.99	2005
BK7780530	Internet Telephone Basics	224	0972805303	$29.99	2003
BK7780532	Voice over Data Networks for Managers	348	097280532X	$49.99	2003
BK7780538	Introduction to SIP IP Telephony Systems	144	0972805389	$14.99	2003
BK7781311	Creating RFPs for IP Telephony Communication Systems	86	193281311X	$19.99	2004
BK7781309	IP Telephony Basics	324	1932813098	$34.99	2004
BK7769430	Introduction to SS7 and IP	56	0974694304	$12.99	2004
IP Television					
BK7781362	Creating RFPs for IP Television Systems	86	1932813624	$19.99	2005
BK7781357	IP Television Directory	154	1932813578	$89.99	2005
BK7781355	Introduction to Data Multicasting	68	1932813551	$14.99	2005
BK7781340	Introduction to Digital Rights Management (DRM)	84	1932813403	$14.99	2005
BK7781351	Introduction to IP Audio	64	1932813519	$14.99	2005
BK7781335	Introduction to IP Television	104	1932813357	$14.99	2005
BK7781330	Introduction to IP Video Servers	68	1932813306	$14.99	2005
BK7781341	Introduction to IP Video	88	1932813411	$14.99	2005
BK7781352	Introduction to Mobile Video	68	1932813527	$14.99	2005
BK7781353	Introduction to MPEG	72	1932813535	$14.99	2005
BK7781342	Introduction to Premises Distribution Networks (PDN)	68	193281342X	$14.99	2005
BK7781354	Introduction to Telephone Company Television (Telco TV)	84	1932813543	$14.99	2005
BK7781344	Introduction to Video on Demand (VOD)	68	1932813446	$14.99	2005
BK7781356	IP Television Basics	308	193281356X	$34.99	2005
BK7781334	IP TV Dictionary	652	1932813349	$39.99	2005
BK7781363	IP Video Basics	280	1932813632	$34.99	2005
Programming					
BK7727875	Wireless Markup Language (WML)	287	0974278750	$34.99	2003
BK7781300	Introduction to xHTML:	58	1932813004	$14.99	2004
Legal and Regulatory					
BK7769433	Practical Patent Strategies Used by Successful Companies	48	0974694339	$14.99	2003
BK7781332	Strategic Patent Planning for Software Companies	58	1932813322	$14.99	2004
BK7780533	Patent or Perish	220	0972805338	$39.95	2003
Telecom					
BK7727872	Introduction to Private Telephone Systems 2nd Edition	86	0974278726	$14.99	2005
BK7727876	Introduction to Public Switched Telephone 2nd Edition	54	0974278769	$14.99	2005
BK7780537	SS7 Basics, 3rd Edition	276	0972805370	$34.99	2003
BK7780535	Telecom Basics, 3rd Edition	354	0972805354	$29.99	2003
BK7727870	Introduction to Transmission Systems	52	097427870X	$14.99	2004
BK7781313	ATM Basics	156	1932813136	$29.99	2004
BK7781302	Introduction to SS7	138	1932813020	$19.99	2004
BK7781345	Introduction to Digital Subscriber Line (DSL)	72	1932813454	$14.99	2005

For a complete list please visit
www.AlthosBooks.com

Althos Publishing Book List
Winter 2005-06

Wireless					
BK7781306	Introduction to GPRS and EDGE	98	1932813063	$14.99	2004
BK7781304	Introduction to GSM	110	1932813047	$14.99	2004
BK7727878	Introduction to Satellite Systems	72	0974278785	$14.99	2005
BK7727879	Introduction to Wireless Systems	536	0974278793	$11.99	2003
BK7769432	Introduction to Mobile Telephone Systems	48	0974694320	$10.99	2003
BK7769435	Introduction to Bluetooth	60	0974694355	$14.99	2004
BK7769436	Introduction to Private Land Mobile Radio	50	0974694363	$14.99	2004
BK7769434	Introduction to 802.11 Wireless LAN (WLAN)	62	0974694347	$14.99	2004
BK7769437	Introduction to Paging Systems	42	0974694371	$14.99	2004
BK7781308	Introduction to EVDO	84	193281308X	$14.99	2004
BK7781305	Introduction to Code Division Multiple Access (CDMA)	100	1932813055	$14.99	2004
BK7781303	Wireless Technology Basics	50	1932813039	$12.99	2004
BK7781312	Introduction to WCDMA	112	1932813128	$14.99	2004
BK7780534	Wireless Systems	536	0972805346	$34.99	2004
BK7769431	Wireless Dictionary	670	0974694312	$39.99	2005
BK7769439	Introduction to Mobile Data	62	0974694398	$14.99	2005
Optical					
BK7781329	Introduction to Optical Communication	132	1932813292	$14.99	2006

Order Form

Phone: 1 919-557-2260
Fax: 1 919-557-2261 **Date:**______________
404 Wake Chapel Rd., Fuquay-Varina, NC 27526 USA

Name:________________________ Title:________________________
Company:____
Shipping Address:____
City:____________________ State:__________ Postal/ Zip:__________
Billing Address:____
City:____________________ State:__________ Postal/ Zip __________
Telephone:____________________ Fax:________________________
Email: ____
Payment (select): VISA ___ AMEX ___ MC ___ Check ___
Credit Card #: ________________________Expiration Date: ______________
Exact Name on Card: ________________________

Qty.	Product ID	ISBN	Title	Price Ea	Total
Book Total:					
Sales Tax (North Carolina Residents please add 7% sales tax)					
Shipping: $5 per book in the USA, $10 per book outside USA (most countries). Lower shipping and rates may be available online.					
Total order:					

For a complete list please visit
www.AlthosBooks.com

Printed in the United States
216195BV00001B/71/A

9 781932 813739